HOW does Science Work?

Exploring Natural and Man-Made Materials

Carol Ballard

PowerKiDS press.

New York

Published in 2008 by The Rosen Publishing Group, Inc.
29 East 21st Street, New York, NY 10010

First Edition

Commissioning Editor: Vicky Brooker
Editors: Laura Milne, Camilla Lloyd
Senior Design Manager: Rosamund Saunders
Design and artwork: Peta Phipps
Commissioned Photography: Philip Wilkins
Consultant: Dr Peter Burrows
Series Consultant: Sally Hewitt

Library of Congress Cataloging-in-Publication Data

Ballard, Carol.
  Exploring natural and man-made materials / Carol Ballard.
     p. cm. — (How does science work?)
  Includes index.
  ISBN 978-1-4042-4278-4 (library binding)
  1. Materials—Juvenile literature. I. Title.

  TA403.2.B345 2008
  620.1'1—dc22
                                        2007031966

Manufactured in China

Acknowledgements:

Cover photograph: A group of colored balloons, Garry Gay/Alamy

Photo credits: Dana Neely/Getty Images 4, Raymond Gehman/Corbis 6, Joanna McCarthy/Getty Images 8, Darrin Jenkins/Alamy 10, Garry Gay/Alamy 11, David Hiser/Getty Images 12, Jean-Philippe Soule/Alamy 13, PicturePress/Getty Images 15, Janine Wiedel/Alamy 19, Marc Schlossman/Getty Images 22, Michael Paul/Corbis 23, Roy Morsch/Corbis 24, gkphotography/Alamy 25, John Humble/Getty Images 28.

The author and publisher would like to thank the models Kodie Briggs, Dylan Chen, Isabelle Li Murphy, and Jessica Li Murphy.

Contents

Words in **bold** can be found in the glossary on p.30

Materials

We use materials to make things. You might be sitting on a wooden chair or reading a book made from paper. Wood and paper are both types of materials.

The clothes you are wearing are all made from materials, such as cotton for T-shirts, wool for sweaters, and leather for shoes.

Buildings are made from many different materials, too.

Some houses have walls made from brick or stone. Some, like this house, have walls made from wood.

TRY THIS! Find different materials

1 Look around the room that you are in.

2 Write a list of ten different things that you can see.

3 Next to each one, write down what it is made from.

You should be able to find many different materials.

Materials come from different places. Some of them come from under the ground, and some come from plants or animals. These are all **natural materials**. Other materials are made by people and are called **man-made materials**.

Materials from the ground

Many materials come from the ground. Some are found near the **surface**, but others are very deep underground.

Metals and **gems** are often buried deep below the ground. Long tunnels called mines are dug to reach them. Metals are used to make things, such as knives, coins, cars, and airplanes. Gems can be used to make jewelry.

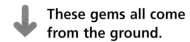

These gems all come from the ground.

TRY THIS! Make your own pot

1 You can make your own clay pot with modeling clay.

2 Roll out long sausages of clay.

3 Wind them around on top of each other. Bake the pot in the oven until it is hard.

You could paint your pot different colors.

! Ask an adult to use the oven for you

Stone is often found nearer the surface of the ground. It is heavy and so it needs strong equipment to dig it out. Stone is used to make things, such as the walls of buildings and statues.

Clay is also found near the surface of the ground. It is easier than stone to dig out because it is softer. Plates, dishes, cups, and vases can all be made from clay.

Rocks and soils

The surface of the Earth is made from rocks. Each type of rock has special **properties**. Rocks were formed millions of years ago. Different rocks were formed in different ways. Some rocks, such as chalk, were made from tiny sea creatures. When the creatures piled up at the bottom of seas and lakes, they were crushed and squashed together to make rocks.

Other rocks, such as granite, were formed deep inside the Earth. It is so hot below the surface of the Earth that the rocks are **liquid**. They cool and become **solid** as they rise to the surface.

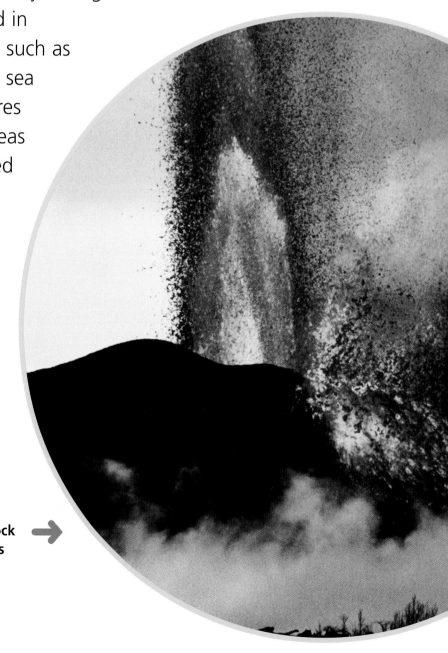

When a volcano erupts, hot liquid rock shoots into the air. The air then cools it down and it becomes solid rock.

TRY THIS! Take a look at soil

1 You will need some soil from outside, a garden sieve, and a bucket to use as a container or a flat surface that you can get dirty.

2 Put some soil into the garden sieve and shake it gently into the container.

3 Some of the soil should go through the sieve into the container, but other twigs and stones might be left in the sieve.

You can see that there are lots of different things in the soil.

Wash your hands after touching soil

Wind, rain, frost, ice, and waves all damage rocks and tiny **fragments** break off. Soil contains these tiny pieces of rock. It also contains dead plants and animals, animal droppings, and tiny creatures, such as worms and beetles.

Materials from plants

Plants provide us with many useful materials. Wood comes from the trunks and branches of trees. It is used to build houses and for parts of buildings, such as beams, floorboards, and doors.

Cotton grows on cotton plants. It is picked and taken to factories where it is spun into long threads. These are woven together to make fabric for things, such as T-shirts, towels, and sheets.

Wood is also used to make furniture such as chairs and tables. ➔

Rubber comes from a sticky juice, called **sap**. Sap can be found inside the trunks of rubber trees. A cut is made in the trunk and the sap trickles into a cup that is strapped to the trunk.

The stems and branches of plants, such as willow, can be dried and woven together to make baskets and furniture.

Rubber can be used to make balloons, balls, elastic, and car tires.

→

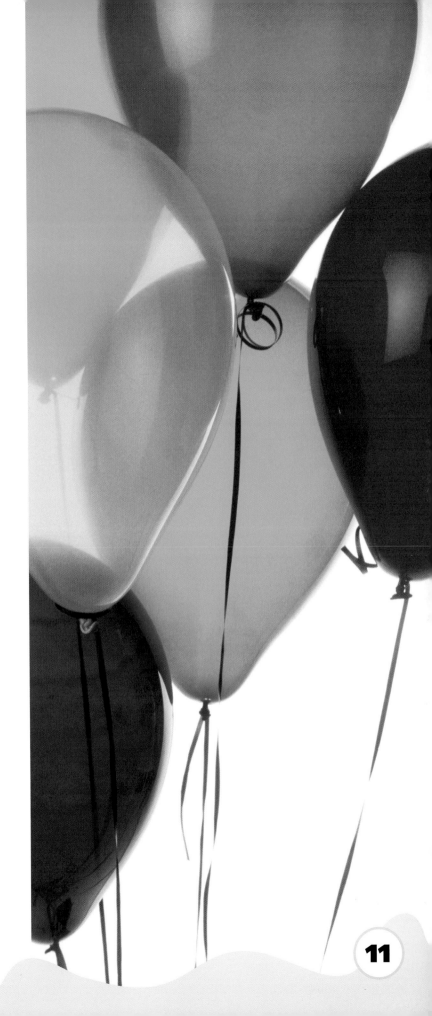

Wow!

The soles of many gym shoes and boots are made from rubber!

Materials from animals

Animals provide us with many useful materials. Wool comes from a sheep's coat. The wool is cut and then spun into threads. Wool is soft and warm and can be used to make things, such as sweaters and scarves, blankets and rugs.

Leather comes from the skins of animals. It is used to make shoes and bags.

In the Arctic, some people wear clothes made from animal fur to keep them warm.

Feathers from birds, such as ducks and geese, are light and soft. They can be used as fillings for pillows, cushions, and comforters. Bright-colored feathers can also be used for decorations.

Silkworms spin **cocoons** of fine silk threads. The threads are woven into fabric.

Silk threads can be spun and woven into fabric to make beautiful clothes, such as these kimonos.

Man-made materials

Some materials are made by people. The material that you start with is called the **raw material**. The new material that is made is called a man-made material.

Plastic is a man-made material. The most important raw material used for making plastic is **oil**. The oil can be separated into different things. Some of these can be used to make plastic.

Wood is the raw material used to make paper. We can use paper for writing, to make books, and for tissues.

Bottles, buckets, and many of the things we use every day are made from plastic.

Very high temperatures are needed to make this big glass container.

The raw materials needed to make glass are sand and limestone. These are mixed together and heated. When the mixture gets very, very hot, it turns into glass. Windows, light bulbs, and drinking glasses are made from glass.

Looking at properties

We can explain what a material is like
by talking about its properties.
Think about how you would describe a
material such as wood. Is it hard or soft?
Is it strong or weak? Is it rough or smooth?
Is it **absorbent** or **waterproof**?

These are just some of the
different properties a
material may have.

**Bricks are hard
but cotton
balls are soft.**

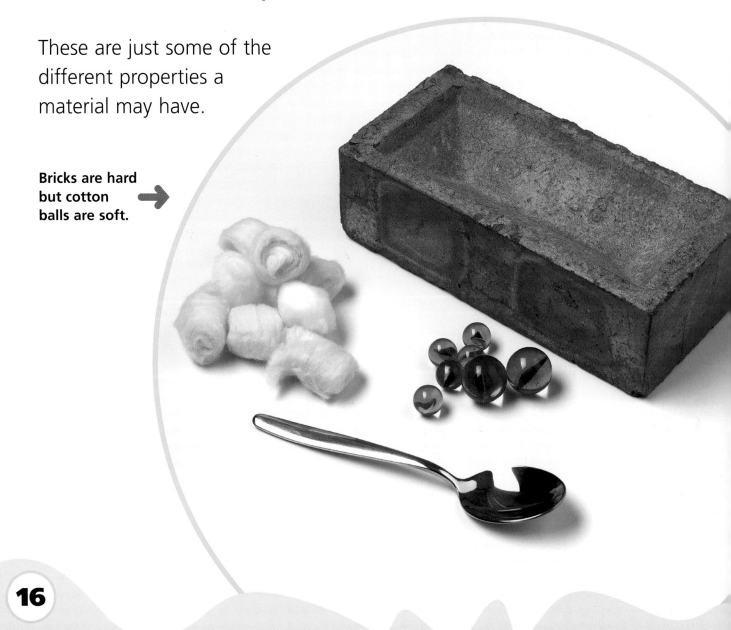

TRY THIS! Test a material's properties

1   You will need to gather some small objects made from different materials, such as wool, coins, and a pebble.

2   List them on one side of a piece of paper.

3   Draw lines to make four more columns down the page.

4   Choose four properties to test the materials for, and write them as headings for the columns.

5   Test the first object for each property, one by one. Touch the object to test whether it is hard or soft. Try pulling it to see if it is elastic. Hold it next to a magnet to test whether or not it is magnetic. After each test, put a ✔ or a ✘ in your table.

6   Then test the rest of the objects for each property until all the boxes are complete.

7   Do any of your materials have exactly the same set of properties?

You can use your chart to help describe each material. For example, a rubber band is soft and elastic.

	hard	soft	elastic	magnetic
wool	✘	✔	✘	✘
rubber band	✘	✔	✔	✘
metal coin				
paper				
pebble				
cushion				

Using materials

It is important to choose materials with the right properties for how they will be used. Can you imagine chairs made from jello and an umbrella made from a towel?

Materials have different properties. This means that some are better for one job than others. An umbrella is supposed to stop rain from getting through, so it needs to be made with a waterproof material.

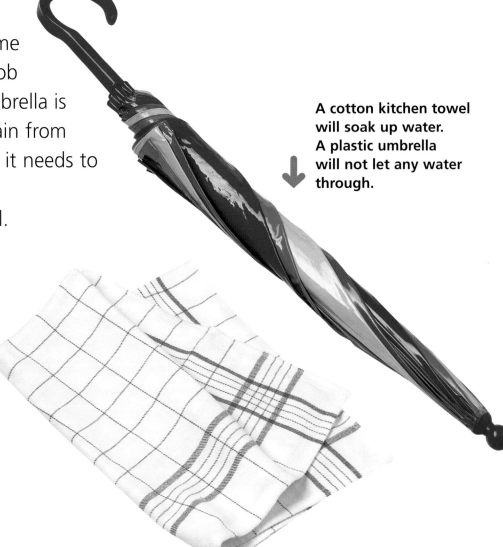

A cotton kitchen towel will soak up water. A plastic umbrella will not let any water through.

A cushion is supposed to be comfortable to sit on, so a soft material is used.

If you look around your bedroom at home or your school classroom, you will see that lots of different materials have been used. Tables and chairs need to be made from something strong and **rigid**, so they are often made from wood, metal, or plastic.

 Are your classroom desks made of wood like these?

Materials for warmth

Materials that do not let heat travel through them are called **thermal insulators**. Wood, wool, and fleece fabrics are all good thermal insulators that keep in warmth.

Warm clothes in winter can help to trap our body heat and keep the cold air out.

Thermal insulators can also keep heat out. Packing a chilled picnic in a thermal picnic bag helps to keep it cool, because it does not let the warm air in.

To keep our bodies warm in cold weather, we can put on hats, scarves, gloves, and sweaters made from wool and fleece materials.

TRY THIS! Find thermal insulators

1 You will need to gather a thermometer and some similar-sized containers made of different materials, such as a metal can, a mug, and a glass.

2 Ask an adult to help you heat some water.

3 Pour the same amount of water into each container.

4 Use a thermometer to measure the temperature of the water.

5 After five minutes, measure the temperature of the water in each container.

6 Repeat after ten and fifteen minutes.

You should find that the water stays hotter in some containers than in others. This is because some materials are better thermal insulators than others.

Be careful with hot water

Materials that let heat travel through them are called **thermal conductors**. Metals and glass are good thermal conductors. Radiators and irons are made of metals, to allow heat to travel through them easily.

Solids and liquids

Some materials change from a liquid to a solid and back again when they are heated or cooled down.

When water is very cold, it turns into solid ice. As it warms up, it **melts** and becomes a liquid. If you put it back in the freezer, it cools. When it gets cold enough, it **freezes** and becomes a solid again.

Ice cream melts quickly on a hot day, but it will turn back to a solid if it is frozen. Remember it is not good to eat ice cream if it has been frozen twice.

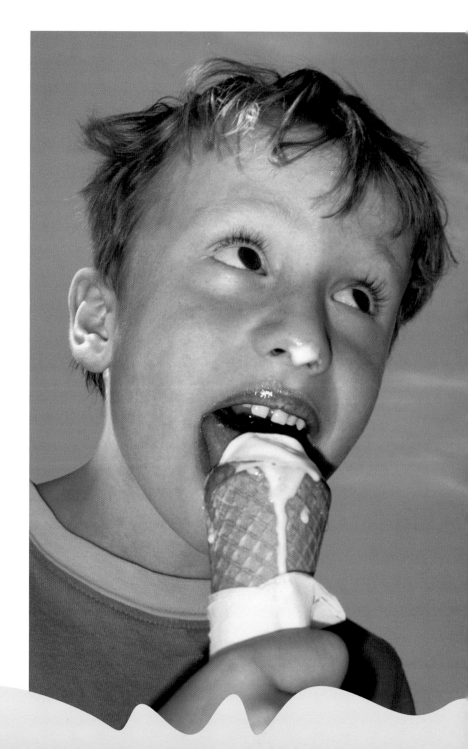

When runny chocolate cools down, it slowly becomes hard and solid again. →

Other materials change as they get hotter or colder, too. For example, butter will melt and become soft and runny on a hot day, but if you put it back in the fridge, it will become hard and solid again.

Wow!

At very high temperatures, even metals and rocks can turn into runny liquids!

Water and steam

What happens to water when it heats up? When water gets hot enough, it **boils** and turns into a **gas** that we call **steam**. The steam disappears into the air.

It is harder to see, but the same thing is happening when a puddle dries up. The water turns into steam and becomes part of the air.

The water in this pan is boiling and is turning into steam.

When steam cools down again, it turns back into liquid water. You can see this happen in a steamy bathroom when water droplets form on cold surfaces such as mirrors and windows.

If you breathe out on a cold day, your breath looks white in the air. This is because the steam in your breath turns back to liquid water when it meets the cold air outside.

Horses' breath turns white when it meets the cold air.

Changed forever

Burning, cooking, and **rusting** all change materials forever. Adding heat to materials makes them change. If we add enough heat, many materials will burn. When a material has burned, you cannot change it back. It turns into soot and ashes, and gases are released into the air. This is what happens when you burn materials such as paper and wood.

When you make toast, you cannot change it back into fresh bread again.

TRY THIS! Change a material forever

1 Squeeze some juice from a lemon into a bowl.

2 Add a teaspoon of baking powder and stir gently.

3 Watch what happens.

The mixture should froth and make bubbles. You cannot get back the lemon juice and baking powder that you started with.

Cooking changes materials forever. For example, when you bake a cake in a hot oven, the heat changes the runny batter into a solid cake. You cannot get the runny cake batter back again.

Recycling materials

Our raw materials will not last forever. This is why we should always try to **recycle** materials.

Recycling means to use things again. A lot of what we throw away can be recycled. This saves raw materials and also means we have less waste.

Glass bottles, jars, and other containers can be heated until they melt. The liquid glass can then be used to make new bottles. Metals can be recycled in the same way.

Plastic bottles and bags can be recycled.

Wow!

Did you know that many fleece jackets are made from recycled plastic bottles?

TRY THIS! Make your own paper

1 You will need a wooden frame, such as an old, unused picture frame, some thin fabric, newspaper, and a bowl of water.

2 Stretch the fabric over the frame and staple it in place.

3 Tear the old newspaper into small pieces and soak them in a bucket of water overnight.

4 Pour most of the water away and squeeze the rest together to make a sticky pulp.

5 Spread this evenly over the fabric and gently press it flat. Cover it with a cloth and leave it to dry.

! Ask an adult to use the stapler for you

When you go back to the dry pulp, you should be able to peel your own recycled piece of paper off of the fabric.

Cardboard and paper can be mixed with water to make a paste. This can then be used to make new cardboard and paper. Recycling paper means we cut down fewer trees.

Glossary

absorbent soaks up liquids

boils when a liquid turns into a gas—for example, liquid water to steam

cocoons silky cases where caterpillars change into adult butterflies or moths

fragments the broken-off parts of something

freezes when a liquid becomes so cold that it turns into a solid—for example, liquid water to solid ice

gas a gas has no shape and can be invisible, air is a mixture of gases

gems stones that come from under the ground

liquid a runny material that takes the shape of its container

man-made materials materials made by people

melts when a solid is heated to become a liquid—for example, solid ice to liquid water

metals hard materials found under the ground

natural materials materials that come from animals, plants, and the Earth

oil a greasy liquid that is found under rocks beneath the Earth's surface

properties what materials are like

raw material a natural material that is used to make a new material

recycle to use again

rigid stiff, not flexible

rusting the slow change from shiny iron to reddish-brown rust

sap juice inside a plant stem

solid a material that keeps its own shape

steam the gas that is made when water is heated

surface the topmost layer of something

thermal conductors materials that heat can travel through

thermal insulators materials that heat cannot travel through

waterproof does not let in water

Further information

Books to read

How We Use... (Using Materials series) by Chris Oxlade, Carol Ballard, et al (Raintree, 2005)

How Do We Use Materials? by Jacqui Bailey (Franklin Watts Ltd., 2005)

Material World series by Claire Llewellyn (Franklin Watts Ltd., 2005)

Materials (Start-up Science) by Claire Llewellyn (Evans Brothers, 2004)

Solids, Liquids, and Gases: From Ice Cubes to Bubbles (Science Answers) by Carol Ballard (Heinemann, 2003)

Web sites to visit

Web Sites
Due to the changing nature of Internet links, PowerKids Press has developed an online list of Web sites related to the subject of this book. This site is regularly updated. Please use this link to access this list:
www.powerkidslinks.com/hdsw/natman

CD Roms to explore

Eyewitness Encyclopedia of Science, Global Software Publishing

I Love Science!, Global Software Publishing

My First Amazing Science Explorer, Global Software Publishing

Index

This story is based on the descriptively enhanced screenplay
developed for "The Animated Stories from the New Testament" video series.
Scripture references have been provided for those readers who
would like to read the story as it is related in the Bible.

Family Entertainment Network, Inc.™
He Is Risen

Matthew 27:50—28:20
Mark 15:39—16:11, 14-20
Luke 23:47-54; 24:1-12, 36-53
John 19:38—20:29; 21:1-17

Stories adapted by:

Sara Clark
Katherine Vawter
Sherry Reeve
Milt Schaffer
Tony Salerno

NEST Publishing
Dallas, Texas

Outside Jerusalem on a hill called "Golgotha" stood three wooden crosses. It was on one of these crosses that Jesus was crucified. Some people did not believe Jesus to be the Messiah. Others, like the high priest, thought that Jesus was trying to become the ruler of Israel. The soldiers put a crown of thorns on Jesus' head to make fun of Him. Then they nailed Him to the cross.

The friends and followers of Jesus gathered around the cross. They were very, very sad. Mary, the mother of Jesus, knelt below the cross in tears. John, one of Jesus' disciples, was there to comfort Mary.

As Jesus died on the cross, the gathering clouds grew darker and darker. A bolt of lightening flashed across the blackened sky.

Jesus' friends looked for safety from the coming storm. Soldiers standing by the cross became nervous.

A soldier, called a centurion, looked up at Jesus. "Beyond doubt, this man was innocent," he said.

Suddenly the earth shook.
People were frightened.
Soldiers were thrown to the ground.

The centurion, looking up to Jesus on the cross, cried out, "Truly this man was the Son of God."

Meanwhile, some of Jesus' disciples, Andrew, Peter, Simon, and James, were meeting at a house in nearby Jerusalem. Andrew stood by a window looking at the stormy sky. Peter sat with his head bowed. Simon walked back and forth. James sat in a corner praying. They were worried and very frightened because Jesus had just been crucified.

James called to Andrew, "Come away from the window, Andrew. They might see you."

Peter asked angrily, "Why should he, James? What does it matter if we live or die?"

Simon walked over to him and said, "It matters, Peter."

"Simon, I denied Him three times!" Peter cried sadly.

Simon put his arm around Peter. "We have to go on, don't you understand?" he said. "We have to fight!"

James remembered what Jesus had taught them. He looked at Simon. "Didn't you learn anything from Him?" he said. "We must never fight." Andrew was still looking out the window. "Even the sky is angry," he said.

Just then there was a knock at the door. It was the secret knock. Two soft, then two loud knocks meant it was a friend. Andrew ran to the door and opened it. His son, Daniel, came into the room quickly.

Daniel said, "Father, I saw Him die."

The disciples were very sad when they heard that. Some of them began to cry. Peter raised his arms to God and cried out, "Why didn't He save himself? He had the power!"

Daniel looked up at Andrew and said, "Father, when He died He looked...glad. The way you look when all the nets are in and it's time to go home."

In a nearby palace, Pontius Pilate, the Roman governor of Judea, sat slumped down on his throne. He looked up to see a man coming toward him. The man was dressed in fine robes and expensive jewelry.

"Procurator Pontius Pilate," the man said as he bowed. "I am Joseph of Arimathea." "Oh, I know you," said Pilate. "You're one of the Sanhedrin. Well, you had your way! Jesus is crucified." Pilate was upset about what had happened to Jesus.

"I ask for the body of Jesus," said Joseph. Joseph had a tomb and wanted to bury Jesus in it.

Pilate leaned forward on his throne. "Can't you even wait until He's dead?"

"But, He is dead, sir," said Joseph sadly. Pilate jumped off his throne. "I don't believe you. Jesus was a strong man!" he shouted.

"Guard!" yelled Pilate. "Where is the centurion in charge of the crucifixion?"

"He's here now, sir," said a guard standing nearby.

The centurion entered the room. He placed his hand over his heart to salute Pilate. Pilate asked about Jesus.

"He is dead, sir," answered the centurion. "We made sure of it before we left the hill."

Pilate dropped his head for a moment. He turned to Joseph, "Why do you want His body?" he asked.

Joseph answered, "I was His friend."

"Friend?" asked the governor. "Now, after He's dead, I meet one of His friends!" Pilate stopped and thought for a moment. Then he said, "Yes, take the body."

Joseph bowed saying, "Thank you, sir." He turned to go.

"Joseph of Arimathea!" called out Pilate. "I would have saved Him if I could." Joseph stood still for a moment. Then without saying a word he left the room. Pilate covered his face with his hands. "So...He's dead," he said sadly.

The centurion turned to Pilate. "Sir, we killed an innocent man," he said.

"Don't you think I know that?" Pilate said. "But I was ordered to keep the peace in this miserable land." Slumping down in his throne again, Pilate said, "He died for peace."

11

Jesus' body had been taken into the tomb. Joseph was preparing Jesus for burial when Nicodemus came to help him.

"Oh, Nicodemus, I'm glad you've come," said Joseph.

"I brought spices to cover Him," said Nicodemus sadly.

Joseph replied, "Yes, enough for a king's burial, my friend."

"That's right," agreed the older man. "A true King."

"Come," said Joseph. "We must wrap His body before the Sabbath." And he helped his friend inside.

Joseph and Nicodemus were members of the Sanhedrin, the highest Jewish court. They were also very good friends - and with good reason. They shared the same faith in Jesus the Messiah. They knew they could be killed for that faith.

They prepared Jesus' body for burial in fresh, white cloth. They laid the body down and prayed over it.

"Farewell, Master," said Joseph. The two men took one last look at their Savior. They stepped outside of the tomb. Then they rolled a heavy stone in front of the entrance to the tomb.

They left the tomb and started down the path toward home. "Come, Nicodemus," Joseph said to his friend. "The Sabbath begins."

Back in the governor's palace in Judea, a group of men were meeting with Pilate. When one of the men stepped forward to speak. Pilate became angry.

"What are you doing here, Caiaphas?" asked Pilate. "Isn't today the Sabbath? I thought you Jews were supposed to stay home and worship, or whatever you do."

Caiaphas, the high priest of the Sanhedrin pointed toward Pilate. "You gave that blasphemer's body to His friends?" he shouted.

"Don't you dare raise your voice to me, Caiaphas!" Pilate yelled. He stood up. His guards drew their swords.

The high priest took one step back. He spoke softly. "Your Excellency, while that impostor was alive He said, 'I am to be raised again after three days.' Tomorrow is the third day. Now His friends will steal His body and claim He rose from the dead. They'll raise rebellions all over Judea. And then Rome will..."

Pilate interrupted Caiaphas. "Oh, don't pretend to be worried about the welfare of Rome. Just tell me what you want!"

"Post a guard at the tomb," said Caiaphas. "Make sure no one steals the body until the third day is past!"

"Indeed, I'll send soldiers to guard it," Pilate agreed. "Caiaphas!" Pilate yelled as the high priest turned to leave. "I did your killing for you. Now, I never want to hear you mention Jesus of Nazareth again!"

Two soldiers and the centurion were sent to Jesus' tomb. The centurion put a notice on the entrance. "This tomb is now sealed by the authority of Rome!" he said.

The centurion gave orders to the soldiers. "You two, keep watch all night. And don't fall asleep! I'll send your replacements at dawn."

When they were alone, one soldier asked, "None of us sleep? After watching all day?"

"Oh, come now," said the other. "It's not like we're working. We are guarding a dead man!"

The two men leaned against the tomb. Soon they were asleep.

Early the next morning, Mary, Martha and Mary Magdalene went to the house where the disciples of Jesus were sleeping. The women knocked. John opened the door and the women entered quickly.

"Wake up and eat," said Martha. "We brought bread and cheese." The men awakened and food was passed around the room.

Thomas asked the women, "Were there any soldiers looking for us?"

"There were soldiers, Thomas," answered Mary Magdalene. "But, I didn't stop to ask them if they were looking for anyone."

Thomas grabbed his coat and headed for the door. "I have to go find my twin brother. He could be arrested in my place!"

John shut the door behind Thomas. Then Mary took a small vase from her bag. "I bought oil to anoint Him," she said.

"They won't let you near Him," said Philip.

"They have soldiers there," Bartholomew agreed.

"Soldiers or not, we're going to the tomb," said Mary Magdalene.

Simon held up a carving knife. "I'd love to show those soldiers what I think of them!" he said, waving the knife around.

"Just eat, Simon," said Martha.

"The true disciples of Jesus aren't killers," said Mary.

"I know," said Simon ashamed.

"But, what are we?" asked Peter. "A bunch of frightened men?"

Young Daniel looked up at Andrew, and asked, "Why can't you still tell people about Jesus, Father?"

James jumped up and said, "Who would believe in Him, now that He's dead?"

"I would," said Mary boldly.

"And shame on you if you don't!" said Martha. She picked up the empty bread baskets and started to leave.

"Wherever He is, He's watching," said Mary Magdalene. "At least we should try to make Him proud of us."

After the women had gone, John turned to the others. "She's right," he said.

"I know it, John," agreed Peter. "I'm just sick of being cooped up in this room. I'd rather die trying to teach His words."

"Yes," said Andrew, "His words..."

The elderly Bartholomew rose from his chair. "Blessed are the pure in heart, for they shall see God," he said.

"Blessed are the meek, for they shall inherit the earth," said Philip as he stood up. John also stood and spoke. "Blessed are those who mourn. For they shall be comforted." Matthew picked up some papyrus and a quill pen. He began to write down the words of Jesus. "His words must never be forgotten," he said.

"Yes, Matthew," agreed James. "They must not die with us."

Meanwhile the two soldiers guarding Jesus' tomb woke up. They stretched and yawned. It was so quiet. Only the sound of a few birds singing could be heard. All of a sudden the ground started to shake. There was a bright light and two angels appeared.

One of the angels was Gabriel. He held out his hand toward the guards and they fell into a deep slee\p. Then the stone that covered the entrance of the tomb rolled away. Inside the tomb was a very bright light.

"Hallelujah!" sang a heavenly choir, "Hallelujah!" Then it was quiet again.

Moments later Mary, Martha and Mary Magdalene arrived at the tomb. They found the stone had been moved from the entrance.

"Oh, no," cried Mary. "What's happened?"

Martha looked inside and saw Jesus' body was gone. "They've taken Him!" she said.

Suddenly two angels stood beside them. The women were frightened. One of the angels, Gabriel, asked, "Why do you look for the living among the dead? He is not here. He is risen."

"How can that be?" asked Mary.

"Remember what He told you?" asked Gabriel. "The Son of man must be delivered into the hands of sinful men, and be crucified, and the third day rise again. Go now and tell the others what you have seen and heard."

The angels disappeared and the women hurried back toward Jerusalem.

When the soldiers guarding the tomb awoke, they looked around. The tomb was empty. "The body's gone!" one soldier shouted. "You saw what happened! The gods have taken Him!"

"Ha!" said the other. "Do you think Pilate is going to believe that? I say we talk to old Caiaphas first. I bet he'll keep us out of trouble in exchange for not telling anybody what happened here this morning."

The three very excited women returned to tell the disciples what they had seen. "He is risen!" Mary exclaimed.

"They've taken Him," said John. He headed for the door.

"Taken away His body?" asked Peter. He ran out the door behind John. Mary Magdalene ran out after them.

Mary looked at the rest of the disciples. She asked, "But don't you understand? He lives!"

Young Daniel turned to Andrew. "Father, is it true?" he asked. "Jesus is alive?"

Andrew looked at his son. "Daniel, we all wish He were alive," he said.

"But we can't let ourselves believe silly rumors," said James. He turned to look out the window.

"It's true!" said Martha. "No matter what you believe, it is true!"

Meanwhile, Peter and John arrived at the open tomb. Peter entered the tomb and looked around. All that was left was a piece of white cloth. He fell to his knees.

"He's gone!" cried John. "They stole the Master's body."

"They couldn't even let His body rest in peace," said Peter angrily. He clutched the cloth and wept.

As Peter and John walked back to Jerusalem. John asked, "Why do they hate Him so much, even when He's dead?"

When Mary Magdalene returned to the empty tomb she fell to her knees and cried. "Oh Master, I believed it when they said You still lived..."

"Why are you weeping?" asked a man standing nearby. "Who is it you're looking for?"

Mary looked toward the man. His face was hidden in a shadow. "If you're the one who took Him away, sir, please tell me where you laid Him," she said.

"Mary!" said the man as he stepped into the morning sun. It was Jesus!

"My Master!" cried Mary Magdalene.

She reached out to touch him, but Jesus stopped her. "Oh, don't touch Me, Mary, for I have not yet ascended to My Father. But go to My brethren, and tell them that I am now ascending to My Father and your Father, My God and your God."

Mary Magdalene ran back to the upper room to tell the disciples about Jesus. "I tell you I saw Him with my own eyes. And He spoke to me!" she said

"I was at the tomb, Mary," said Peter. "And He wasn't there!"

"Why would He show Himself to you, Mary, and not to us?" asked John.

"Well, maybe He took pity on me because I was crying," she answered. She was sad that they did not believe her.

Martha put her arm around Mary Magdalene. "They won't believe you, Mary. They didn't believe us, either," said Martha. "We've tried all day to persuade them."

"We don't think you're lying," said Peter.

"Just imagining things," added Philip.

"But you should believe us," said the other Mary to the disciples.

"With all my heart I wish I could," said Bartholomew.

Suddenly, the room was filled with light. Jesus stood before them. "Peace be with you," said the Savior.

Everyone was afraid. But Peter spoke out. "Is it You?"

"Look at My hands and My feet," Jesus answered. "Touch Me and see. For a spirit does not have flesh and bones as you see Me have."

John reached out and touched Jesus' hand. One by one each disciple fell to his knees before Jesus. They were sorry they had not believed that He had risen from the dead.

29

In the council chamber of the Sanhedrin, the high priest, Caiaphas was listening to the story of the two soldiers who had been guarding the tomb. "Angels?" asked the priest. "Nonsense. They must have bribed you."

"Nobody bribed me!" shouted the first soldier.

"How much?" growled the priest. "How much money will it take to make you tell the truth? That they came and stole the body."

"Nobody stole anything," said the soldier.

"We'll tell the truth," said the other. "We were asleep when the body disappeared."

Caiaphas had a bag of silver coins. He asked the soldiers, "Is this enough to help you both remember that story and no other?"

"Well, I don't know..." said one soldier. "Pilate will be very angry with us for sleeping on duty."

"Oh, I'll protect you from Pilate!" yelled the priest. He threw the money at them.

"Nice doing business with you," said one of the soldiers as they left the room.

"We already paid thirty pieces of silver to have this Jesus dead," muttered Caiaphas to himself. "Now we pay even more to keep Him dead!"

The apostle Thomas had not been with the disciples when Jesus had appeared before. They were trying to tell him about the risen Lord. "And you expect me to believe this?" asked Thomas. "I'm gone for a week. And I come back and you've all lost your minds!"

"We saw Him, Thomas," said John. "We touched Him."

"I'm sorry, John," said the doubting Thomas. "But, I'm a reasonable man. If I don't see it with my own eyes, then it isn't...isn't..." Suddenly, the room filled with a bright light. Thomas' eyes opened wide.

"Peace be with you, Thomas," said Jesus. He came up to Thomas. "See My hands and My feet. Touch Me. Doubt no longer, Thomas, but believe."

Thomas fell at Christ's feet. "My Lord, my God," he said.

"Because you saw Me, you believe," said Jesus. "Blessed are those who have not seen Me, and yet believe."

Back in Pilate's Hall, the centurion stood before the governor. "Sir, the story flies through Jerusalem. Jesus is risen, they say. His apostles and several women have seen Him and touched Him."

"Then arrest them!" shouted the governor. "Arrest all of them! And stop them from telling those lies."

"We can't, sir," said the centurion. "They've escaped into Galilee."

"If your soldiers had done their duty..." said Pilate.

"They did their duty!" the centurion said. "But all the Roman Legions could not have held Jesus of Nazareth."

Pilate looked up at the centurion. "You really believe it!" said Pilate, laughing. "You believe He was resurrected!"

The centurion looked at Pilate. He didn't say anything.

Then Pilate turned his back to the centurion. "Well, I don't believe it!" he said. "I did not order the killing of the Son of God!"

Out on the Sea of Galilee, Peter, James, John, Andrew, and Daniel pulled their fishing net from the water. It was empty. Not a fish in sight.

"I give up," said Peter. "Empty nets all night!"

Young Daniel looked out toward shore. "Look, father!" said Daniel. "There's a man on shore watching us."

The man called out to the boat, "Friends, have you caught anything?" "No, nothing," Andrew shouted back.

"Cast the net on the right side of the ship, and you'll make a catch," the stranger said.

"On the right side?" asked Peter. "We cast on that side only an hour ago. What difference does it make?"

"Come on Peter," said John. "It's worth a try."

So, the men threw the nets into the water on the right side of the ship. Then they started to pull the nets back in.

"Pull, James!" shouted Andrew.

"I am pulling, Andrew!" James shouted back.

When Daniel looked over the edge of the boat he saw a net full of fish. "Father!" the boy laughed. "There's too many fish!"

"I...I've never seen so many fish, all at one time!" said Andrew.

"Not a fish all night, and now..." Peter said. He looked out toward the stranger. Suddenly he knew. It was no stranger. It was Jesus! Peter dove in the water and swam to shore.

Andrew called out. "Quickly! Raise the sail. Let's get ashore!"

39

That evening the men sat together on the shore. They cooked their fish and talked. As the fire warmed them, Jesus talked with the disciples. He asked Peter, "Simon Peter, do you love Me more than these?"

Peter answered, "Yes, Lord. You know that I love You."

"Then feed My lambs," said Jesus. He placed his hand on young Daniel's shoulder.

"I will," promised Peter.

"Simon Peter, do you love Me?" asked Jesus.

"Yes, Lord, you know that I love You," Peter said again.

"Then feed My sheep," said Jesus. Peter nodded.

"Simon Peter, do you love Me?" Jesus asked again.

Once again Peter answered. "Master, you know everything. And you know that I love You."

Jesus said one last time, "Feed My sheep."

Later the disciples returned to Jerusalem. For forty days Jesus taught them. Then it came time for Him to leave and return to heaven. Hundreds of his followers gathered in a field outside of Jerusalem to hear Him one last time.

Jesus said. "Go and teach all nations, baptizing them in the name of the Father, and the Son, and the Holy Ghost. Teach them to observe all that I have commanded you. And, lo, I am with you always, even to the end of time."

And with that, the Lord Jesus went up into the heavens.

"Ye men of Galilee," the angel Gabriel called out to those gathered there. "Why do you stand there gazing into heaven? This same Jesus which is taken from you into heaven shall come again, in like manner as you have seen him go."

Young Daniel stood there watching. Tears were rolling down his cheeks. He called out to his Savior, "Come quickly, Lord Jesus! I'll always be watching for You!"

THE END

HE GAVE HIS LIFE FOR ME

Lyrics by
JULIE DE AZEVEDO
and
LEX DE AZEVEDO

Music by
LEX DE AZEVEDO
Arranged by
PAUL FISCHER

48